SpringerBriefs in Cognitive Computation

Volume 6

Editor-in-chief

Amir Hussain, Stirling, UK

About the Series

SpringerBriefs in Cognitive Computation are an exciting new series of slim high-quality publications of cutting-edge research and practical applications covering the whole spectrum of multi-disciplinary fields encompassed by the emerging discipline of Cognitive Computation. The Series aims to bridge the existing gap between life sciences, social sciences, engineering, physical and mathematical sciences, and humanities.

The broad scope of Cognitive Computation covers basic and applied work involving bio-inspired computational, theoretical, experimental and integrative accounts of all aspects of natural and artificial cognitive systems, including: perception, action, attention, learning and memory, decision making, language processing, communication, reasoning, problem solving, and consciousness.

More information about this series at http://www.springer.com/series/10374

Maria Teresa Riviello · Anna Esposito

On the Perception of Dynamic Emotional Expressions: A Cross-Cultural Comparison

Springer

Maria Teresa Riviello
Department of Psychology
Seconda Università di Napoli
Caserta, Caserta
Italy

and

International Institute for Advanced Scientific Studies (IIASS)
Vietri sul Mare
Italy

Anna Esposito
Department of Psychology
Seconda Università di Napoli
Caserta, Caserta
Italy

and

International Institute for Advanced Scientific Studies (IIASS)
Vietri sul Mare
Italy

ISSN 2212-6023 ISSN 2212-6031 (electronic)
SpringerBriefs in Cognitive Computation
ISBN 978-94-024-0885-0 ISBN 978-94-024-0887-4 (eBook)
DOI 10.1007/978-94-024-0887-4

Library of Congress Control Number: 2016944161

Printed on acid-free paper

This Springer imprint is published by Springer Nature
The registered company is Springer Science+Business Media B.V. Dordrecht

Preface

In social interactions, the communication of emotions is an example of a multimodal transfer of information: the emotional cues are conveyed and perceived simultaneously through several modalities, mainly represented by facial and vocal expressions.

The coding and processing of emotional signals is a challenging objective in the context of Human Computer Interaction (HCI). The research aims at defining methods and procedures capable of automatically identifying human emotional states exploiting the multimodal nature of emotional expressions. The prime goal is allowing the implementation of emotionally interactive dialogues systems, wearable computing, and embodied conversational agents, in order to ease and simplify user access to both nearby and remote telecommunication services.

Most of the research on emotions and related perceptual cues to infer them has focused on three separately expressive domains: facial expressions, voice, and body movements. In addition, studies investigating on emotional facial expressions have mostly exploited static images, neglecting the dynamicity of facial movements. This work is devoted to explore the power of visual and vocal channels, in conveying the emotional cues exploiting realistic, dynamic, and mutually related emotional vocal and facial stimuli. The main goal is to investigate if one channel is more effective than the other for inferring emotional information, and if this effectiveness is affected by the cultural context generated by the spoken language when close cultures are considered. For this purpose we set up perceptual experiments devoted to assess the amount of emotional content that participants from different Western cultures and languages—Italian, American, French, Hungarian, and Lithuanian—can infer from audio, mute video, and audio–video emotional expressions of happiness, fear, anger, irony, surprise, and sadness. The emotional stimuli were obtained from movie scenes relating to two different cultural contexts and played in two different languages: American English (as a globally spread language) and Italian (as a country-specific language).

The introduction to the reported work, the definition of the topics, and the rationale are illustrated in Chap. 1. The detailed description of materials, participants, and procedures exploited in the proposed research experiments, as well as data analyses is the content of Chap. 2. The results are shown and discussed separately for the Italian (Chap. 3), American (Chap. 4), French (Chap. 5), Hungarian (Chap. 6), and Lithuanian (Chap. 7) participants. A further elaboration of the collected data illustrating the cross-cultural comparison is contained in Chap. 8. Chapter 9 presents conclusions and comments.

Caserta and Vietri sul Mare, Italy Maria Teresa Riviello
Caserta and Vietri sul Mare, Italy Anna Esposito

Contents

Chapter 1
Introduction

Abstract The coding and processing of emotional signals are relevant objectives in the context of Human Computer Interaction, where research aims at defining methods and procedures capable of automatically identifying human emotional states exploiting the multimodal nature of emotional expressions. Since the above research line requires the analysis of verbal and nonverbal interactional emotional cues, further efforts are worth to be made in order to go into the details of the human emotional decoding process. This will allow clarifying the cognitive mechanisms underlying the human perception of emotional expressions, as well as the identification of individual and cross-cultural differences among such perceptual processes. An overview of the works in this field, as well as, the debate on the universality of emotional expressions represents the introduction and the rationale of the current research work. It is an attempt to define and identify distinctive emotional features from multi-modal and cross-cultural dynamic signals and contribute to the development of new methodologies and mathematical models for the automatic implementation of naturally human-like communication interfaces.

Keywords Emotional signals · Decoding perceptual processes · Automatic emotion recognition · Multimodal expressions · Cross-cultural emotional experiments · Human-computer interaction · Emotional dynamic stimuli · Auditory emotional information · Visual emotional information

Emotional expressions play a vital role in human interactions. In everyday life, expressions of emotions refer to the way people communicate experience and influence relationships. During human interactions, this information is transmitted and perceived trough several sensorial channels simultaneously exploiting both the verbal and nonverbal (gesture, gaze, tonal expression) communication modes.

The role of emotional communication in the context of Human Computer Interaction (HCI) is becoming even more relevant and challenging. HCI for affective systems embraces theories from a wide range of domains and disciplines such as psychology and sociology, robotics, computer science, engineering, and system's design. It is relevant to a diverse set of application areas, from teaching,

M.T. Riviello and A. Esposito, *On the Perception of Dynamic Emotional Expressions: A Cross-Cultural Comparison*, SpringerBriefs in Cognitive Computation 6, DOI 10.1007/978-94-024-0887-4_1

learning, office applications, to entertainment technology, therapeutic applications, advertising, and product design. Continual areas of interest within research include the recognition, as well as, the synthesis of affect and emotion in the face, body, and speech.

Given the complexity and multimodal nature of the phenomenon, there has been a branching of engineering approaches toward the improvement and development of video-audio processing, recognition and synthesis techniques with the goal to develop new methodologies for recognizing emotional states from faces (Chickerur and Joshi 2015; Valstar et al. 2014, among many others), speech (Poria et al. 2015a; Shaikh Nilofer et al. 2015, among many others) and body movements (Wang et al. 2016).

Humans detect, recognize, differentiate and identify emotions in various situations and contexts with a multimodal approach. Actually, the emotional signal may be not under clear condition as a posed and well-orientated face and/or a noiseless and unambiguous speech; in addition, numerous factors affect the detection and interpretation of emotional expressions. Among these cultural characteristics: it seems that the experience and familiarity with faces from other cultures affect the recognition of emotional expressions (Yan et al. 2016; de Gelder and Huis in ‘t Veld 2016), and that cultural variability in the accuracy of emotion decoding process depends on differences in language (Riviello et al. 2014).

On these premises, this work attempts to go into the mechanism underlying the human perception of emotional expressions through perceptual experiments that analyze the ability of recognizing emotions from realistic and dynamic data, created to allow an actual comparison between the effectiveness of visual and vocal cues in conveying emotional expressions. The databases of stimuli used in the experiments, in fact, consists in mutually related audio, mute video, and audio-video stimuli, extracted from live recording movie scenes. The databases of emotional stimuli relate to American and Italian context and language and the participants to the experiments belong to five different Western culture and languages. Thus, the conducted experiments allow exploring the possible variation in inferring emotions across similar cultures, highlighting the role of the familiarity with the cultural context and language.

1.1 The Expressions of Emotions

Emotional expressions can be defined as perceptible verbal and non-verbal behaviors that communicate an internal affective state.

Over the last two centuries, researchers have proposed different and often competing models explaining emotions and emotional expressions, however, all theorists in emotion agree that all humans principally express emotions with their voices (Bänziger et al. 2015) and faces (Guarnera et al. 2015).

1.1.1 The Facial Expression of Emotions

After Darwin (1965), who was the first to propose that facial expressions are indicator of emotions, rudiments of appropriate behaviors with important adaptive and survival functions, Tomkins (1982) perpetuated Darwin's fundamental assumption of a biologically based emotional mechanism clearly implying inter-cultural universality of the emotion process. To these, Tomkins added the idea of a small, fixed number of discrete ("basic") emotions. According to Tomkins's theory, each basic emotion can vary in intensity and consists of a single brain process (an "affect program"), whose triggering produces all the various manifestations (components) of the emotion, including its facial expression.

In this research line, Ekman (1992a, b) and Izard (1994) described how the contraction of each facial muscle (singly and in combination with other muscles) changes the appearance of the faces when a certain emotion is experienced. Anatomically based coding systems (the Facial Action Coding System, (FACS) (Ekman et al. 2002), the Maximally Discriminative Facial Movement Coding System (MAX) and AFFEX (Izard 1979; Izard et al. 1983) were then developed for measuring facial behaviors and identifying through facial muscle movements related emotional expressions. The three proposed coding systems are different both in the primary emotions they include (only six for Ekman—happiness, sadness, anger, fear, surprise, and disgust; nine for Izard) and in the set of the associated facial muscle movements they assign to some of these emotional states.

On the other side, several authors debated the idea of the universality of the emotions' expressions. White (1993) and Fridlund (1994) reckoned that any behavior, and therefore also expressions of emotions, are learned and thus they vary across cultures. Recent studies attempt to support this idea (see Gendron et al. 2014, 2015, among many others).

Recent theoretical models have attempted to account for both universality and cultural variations by specifying which particular aspects of emotions show similarities and differences across cultural boundaries (Shao et al. 2015).

1.1.2 The Vocal Expression of Emotions

Emotional information in speech is transmitted through explicit (linguistic) and implicit (paralinguistic) messages, exploiting prosodic and acoustic speech features. Emotions, in fact, affect vocalizations, since they produce changes in respiration, phonation, and articulation (Patel and Scherer 2013).

Emotional information from speech is derived from perceptual cues (prosodic aspects of the speech) such as loudness, pitch, and timing, which in turn are acoustically expressed in changes of variables such as amplitude (which is the measure of sound pressure), and sound intensity (which is the measure of sound energy at different frequencies) for loudness; the fundamental frequency (F0) of the signal (measured in Hz), and some derived measures on F0 for pitch; speaking rate, utterance and syllable lengths, empty and filled pauses for timing.

The identification of the acoustic features that convey the vocal expression of emotions is based primarily on the analysis of emotional speech, which allowed identifying, through processes of normalization with respect to non-emotional speech, the ranges of variation of the different parameters. Based on the values of these parameters, the general characteristics that convey emotions in speech have been identified as well as their hierarchy of importance, which sees in the first position the F0 values, followed by intensity and duration values (Arias et al. 2014; Lin et al. 2012).

These characteristics were found to be sufficiently shared cross-linguistically (Bryant and Barrett 2008). This consistency of data across languages, rather than an exclusive universal tendency in the vocal expression of primary emotions, can be also a methodological bias originated by of the data collection procedure. Emotional speech has been so far produced by actors on the basis on constrained scenarios and/or lexical labels, generating stereotypical and prototypical vocal forms, usually characterized by a high activation (Drolet et al. 2014).

1.2 The Expression of Emotions in Human Computer Interaction

In a paper from 1995, Rosalind Picard introduced "affective computing" as a term to describe the computing that relates to, arises from, or influences emotions. According to this, machines should interpret the emotional state of humans and adapt their functional behavior to them, giving appropriate responses for specific emotions. A computer that could decode and produce affective responses has the potential for significant human acceptance and improvement of its interactive capabilities.

This line of inquiry has widespread implications for Human Computer Interaction (HCI), ranging from better educational software to improved computer-mediate communications.

In HCI for affective systems, the research objectives are to identify methods and procedures capable of automatically identifying and modeling human emotional states, exploiting the multimodal nature of emotions. This requires the consideration of several key aspects, such as the development and the integration of algorithms and procedures for the recognition of emotional states from speech and facial expressions, for implementing friendly human/computer interactions.

Principally, research in this field approaches the problem of recognizing emotional facial expressions exploiting static images as the ones contained in Facial Action Coding System, FACS) by Ekman et al. (2002), Zhang et al. (2015), the Japanese Female Facial Expression (JAFFE) database, by Kamachi et al. (1998), Thuseethan and Kuhanesan (2014), the ORL Database of Faces by Samaria and

Harter (1994), Roychowdhury and Emmons (2015), or, more recently, the Multimedia Understanding Group (MUG) facial expression database by Aifanti et al. (2010), Ghimire et al. (2015), among many others, as reported in http://www.face-rec.org/databases/. These databases mostly contain static images, which usually capture the apex of the emotional expressions, i.e. the instant at which the indicators of the expressed emotion are most marked, neglecting that in daily experience, emotional states are intrinsically dynamic processes and associated facial expressions vary along time.

Research has also been conducted on the extraction of emotional expressions from video sequences. Most works in this area exploit video database developed from subjects making expressions on demand, such as the Cohn-Kanade facial expression database by Kanade and Cohn (2000), Zheng et al. (2015), neglecting the importance of the context, or under strict lab conditions, as the recent BP4 D-Spontaneous Facial expressions database (Zhang et al. 2014).

On the other side, most of the existing emotional audio databases, consist of audio stimuli recorded under studio conditions, such as the Berlin Database of Emotional Speech (BDES), by Burkhardt et al. (2005), Vásquez-Correa et al. (2016), or taken from TV, as the VAM database by Grimm et al. (2008), Oflazoglu and Yildirim (2013). In both the cases the emotional voice do not possess genuine emotional characteristics, because they are posed in the first case, and biased by the presence of cameras and the consciousness to be on the spot, in the second one.

By reading the same sentence or watching the same face with different simulated emotional states, it should be possible to allow principled analysis of speech and faces and comparisons of the changing acoustic and visual features. Since such emotions are acted, without a context, or brought to extremes, it is questionable whether they authentically represent the characteristics of speech and faces made by ordinary people when they spontaneously experience similar emotions. Moreover, in face-to-face communication, humans employ facial and vocal modality alone or using one to complement and enhance another. The roles of multiple modalities and their interaction remain to be quantified, in order to further improve emotion recognition systems (Poria et al. 2015b).

References

Aifanti, N., Papachristou, C., & Delopoulos, A. (2010). The mug facial expression database. In *2010 11th International Workshop on Image Analysis for Multimedia Interactive Services (WIAMIS)* (pp. 1–4). IEEE.

Arias, J. P., Busso, C., & Yoma, N. B. (2014). Shape-based modeling of the fundamental frequency contour for emotion detection in speech. *Computer Speech & Language, Computer Speech and Language, 28*, 278–294 (Elsevier).

Bänziger T., Hosoya, G., & Scherer K. R. (2015). Path models of vocal emotion communication. *PLoS ONE, 10*(9), e0136675. doi:10.1371/journal.pone.0136675. Editor: David Reby, University of Sussex, UK.

Bryant, G., & Barrett, H. (2008). Vocal emotion recognition across disparate cultures. *Journal of Cognition and Culture, 8*, 135–148.

Burkhardt, F., Paeschke, A., Rolfes, M., Sendlmeier, W., & Weiss, B. (2005). A database of German emotional speech. In *Proceedings of Interspeech, ISCA, Lisbon, Portugal.*

Chickerur, S., & Joshi, K. (2015). 3D face model dataset: Automatic detection of facial expressions and emotions for educational environments. *British Journal of Educational Technology, 46*, 1028–1037. doi:10.1111/bjet.12325.

Darwin, C. (1965). *The expression of the emotions in man and the animals*. Chicago, IL: Chicago Press.

de Gelder, B., & Huis in 't Veld, E. M. J. (2016). Cultural differences in emotional expressions and body language. In J. Chiao et al. (Eds.), *Handbook of Cultural Neuroscience*.

Drolet, M., Schubotz, R. I., & Fischer J. (2014). Recognizing the authenticity of emotional expressions: F0 contour matters when you need to know. *Frontiers, Human Neuroscience, 8*. doi:10.3389/fnhum.2014.00144.

Ekman, P. (1992a). Facial expression of emotion: New findings, new questions. *Psychological Science, 3*, 34–38.

Ekman, P. (1992b). An argument for basic emotions. *Cognition and Emotion, 6*, 169–200.

Ekman, P., Friesen, W. V., & Hager, J. C. (2002). *The facial action coding system*. Salt Lake City: Research Nexus.

Fridlund, A. J. (1994). *Human facial expressions: An evolutionary view*. San Diego, CA: Academic press.

Gendron, M., Roberson, D., van der Vyver, J., & Barrett, L. (2014). Cultural relativity in perceiving emotions from vocalizations. *Psychological Science, 4*, 911–920. doi:10.11770956797613517239/0956797613517239.

Gendron, M., Roberson, D., & Barrett, L. F. (2015). Cultural variation in emotion perception is real a response to Sauter, Eisner, Ekman, and Scott. *Psychological Science*. doi:10.1177/0956797614566659.

Ghimire, D., Lee, J., Li, Z. N., Jeong, S., Park, S. H., & Choi, H. S. (2015). Recognition of facial expressions based on tracking and selection of discriminative geometric features. *International Journal of Multimedia and Ubiquitous Engineering, 10*(3), 35–44.

Grimm M., Kroschel, K., & Narayanan, S. (2008). The Vera am Mittag German audio-visual emotional speech database. In *IEEE International conference on multimedia and expo (ICME)*. Hannover, Germany: IEEE.

Guarnera, M., Hichy, Z., Cascio, M., & Carrubba, S. (2015). Facial expressions and ability to recognize emotions from eyes or mouth in children. *Europe's Journal of Psychology*, North America.

Izard, C. E. (1979). The maximally discriminative facial movement coding system (MAX). Unpublished manuscript. Available from Instructional Resource Center, University of Delaware.

Izard, C. E. (1994). Innate and universal facial expressions: Evidence from developmental and cross-cultural research. *Psychological Bulletin, 115*, 288–299.

Izard, C. E., Dougherty, L. M., & Hembree, E. A. (1983). A system for identifying affect expressions by holistic judgments. Unpublished manuscript. Available from Instructional Resource Center, University of Delaware.

Kamachi, M., Lyons, M., & Gyoba, J. (1998). Japanese Female Facial Expression Database, Psychology Department in Kyushu University. http://www.kasrl.org/jaffe.html.

Kanade, T., & Cohn, J. F. (2000). Comprehensive database for facial expression analysis. In *IEEE Proceedings of the Fourth International Conference on Automatic Face and Gestures Recognition, Grenoble, France* (pp. 46–53) March, 2000. ISBN: 0-7695-0580-5.

Lin, J. C., Wu, C. H., & Wei, W. L. (2012). Error weighted semi-coupled hidden Markov model for audio-visual emotion recognition. *IEEE Transactions on Multimedia, 14*, 142–156.

Oflazoglu, C., & Yildirim, S. (2013). Recognizing emotion from Turkish speech using acoustic features. *EURASIP, Journal on Audio, Speech, and Music*. doi:10.1186/1687-4722-2013-26 (Springer).

Patel, S., & Scherer, K. R. (2013). Vocal Behaviour. In J. A. Hall, M. L. Knapp (Eds.), *Handbook of nonverbal communication* (pp. 167–204). Berlin: Mouton-De Gruyter.

Picard, R. (1995). Affective Computing. Technical Report 321, MIT Media Laboratorium Perceptual Computing Section, Cambridge, Massachusetts.

Poria, S., Cambria, E., Gelbukh, A., Bisio, F., & Hussain, A. (2015a). Sentiment big data flow analysis by means of dynamic linguistic patterns. *IEEE Computational Intelligence Magazine, 10*(4), 26–36 (ISI-SCI IF: 2.6).

Poria, S., Cambria, E., Hussain, A., & Huang, G. (2015b). Towards an intelligent framework for multimodal affective data analysis, (Elsevier) *Neural Networks, 63*, 104–116. doi:10.1016/j.neunet.2014.10.005, March 2015 (ISI-SCI IF: 2.7).

Riviello, M. T., Maskeliunas, R., Kruminiene, J., & Esposito, A. (2014). Language and gender effect in decoding emotional information: A Study on Lithuanian Subjects. In S. Bassis et al. (Eds.), *Recent Advances of Neural Network Models and Application, Smart Innovation, Systems and Technologies* (Vol. 26, pp. 353–361). Switzerland: Springer International Publishing. ISBN 978-3-319-04128-5.

Roychowdhury, S., & Emmons, M. (2015). A survey of the trends in facial and expression recognition databases and methods. *International Journal of Computer Science & Engineering Survey, 6*, 1–19. doi:10.5121/ijcses.2015.6501.

Samaria, F., & Harter, A. (1994). *The ORL Database of Faces*. Cambridge: Cambridge University Press. http://www.cl.cam.ac.uk/research/dtg/attarchive/facedatabase.htm.

Shaikh Nilofer R. A., Gadhe, R. P., Deshmukh, R. R., Waghmare, V. B., & Shrishrimal, P. P. (2015). Automatic emotion recognition from speech signals: A review. *International Journal of Scientific & Engineering Research, 6*(4). ISSN 2229-5518.

Shao, B., Doucet, L., & Caruso, D. R. (2015). Universality versus Cultural Specificity of Three Emotion Domains. Some Evidence Based on the Cascading Model of Emotional Intelligence. *Journal of Cross-Cultural Psychology, 46*(2), 229.

Tomkins, S. S. (1982). *Affect, imagery, consciousness: Vol. 3. Cognition and affect*. New York: Springer.

Thuseethan, S., & Kuhanesan, S. (2014). Eigenface based recognition of emotion variant faces. *Computer Engineering and Intelligent Systems, 5*(7), 31–37. ISSN 2222-1719 (Paper) ISSN 2222-2863 (Online).

Valstar, M., Schuller, B., Smith, K., Almaev, T., Eyben, F., Krajewski, J., Cowie, R., & Pantic, M. (2014). 3D dimensional affect and depression recognition challenge. In *International Workshop on Audio/Visual Emotion Challenge* (pp 3–10).

Vásquez-Correa, J. C., Orozco-Arroyave, J. R., Arias-Londoñ, J. D., Vargas-Bonilla, J. F., & Nöth, E. (2016). Non-linear Dynamics Characterization from Wavelet Packet Transform for Automatic Recognition of Emotional Speech. Recent Advances in Nonlinear Speech Processing Vol. 48, Smart Innovation, Systems and Technologies, pp 199–207.

Wang, W., Enescu, V., & Sahli, H. (2016). Adaptive Real-Time Emotion Recognition from Body Movements. *ACM Transaction on Interactive Intelligent System (TiiS)*, 5(4), 18. New York, USA.

White, G. M. (1993). Emotion inside out the anthropology of affect. In M. Haviland, & J. M. Lewis (Eds.), *Handbook of Emotion* (pp. 29–40). New York, NY: Guilford Press.

Yan, X., Andrews, T. J., & Young, A. W. (2016). Cultural similarities and differences in perceiving and recognizing facial expressions of basic emotions. *Journal of Experimental Psychology: Human Perception and Performance, 42*(3), 423–440.

Zhang, X., Yin, L., Cohn, J. F., Canavan, S., Reale, M., Horowitz A., Liu, P., & Girard, J. M. (2014). BP4D-Spontaneous: A high-resolution spontaneous 3D dynamic facial expression database. *Image and Vision Computing, 32*(10), 692–706 (October 2014). doi:10.1016/j.imavis.2014.06.002.

Zhang, L., Jiang, M., Farid, D., & Hossain, M. (2015). Intelligent facial emotion recognition and semantic-based topic detection for a humanoid robot. *Expert Systems with Applications, 40*(13), 5160–5168.

Zheng, N., Qi, L., & Guan, L. (2015). Multiple-manifolds Discriminant Analysis for Facial Expression Recognition from Local Patches Set. Multimodal Pattern Recognition of Social Signals in Human-Computer-Interaction. Vol. 8869, Lecture Notes in Computer Science pp. 26–33.

Chapter 2
The Proposed Research Work

Abstract This chapter describes in details the materials, procedures, participants, testing, and data analyses exploited for the cross-modal and cross-cultural experiments on the perception of emotions narrated in this book. Two databases consisting of realistic, dynamic, and mutual related visual and vocal emotional information are described. The databases were constructed to allow an actual comparison between the effectiveness of visual and vocal cues in conveying emotional expressions. In order to investigate if the ability to recognize emotional expressions as a function of the channel is also affected by the cultural context and in particular by the language, the stimuli were extracted from American (as a global spread language) and Italian (as a country specific language) live recording movies. The experiments involved participants from five different Western cultures and languages to explore possible variations in inferring emotions across similar cultures, highlighting the role of the familiarity with the cultural context and language. The data collected were analyzed computing different repeated ANOVA measurements with the goal to reveal the effect of the communication channel, and both stimuli and participants "cultural context and language" on the recognition of the emotional expressions, as well as, to allow a cross-cultural comparison of the results.

Keywords Database of emotional stimuli · Cross-cultural comparisons · Emotion recognition · Cultural effect on emotions perception · Language familiarity · Cultural specificity · Data analysis · Cultural context · Auditory emotional information · Visual emotional information

This work focuses on the cross-modal and cross-cultural analysis of emotional data in the attempt to clarify mechanisms underlying the human perception of emotional expressions, as well as identifying close cross-cultural differences among such perceptual processes (Esposito 2007, 2009).

For this purpose, perceptual experiments exploiting two multimodal databases of realistic, dynamic and mutually related vocal and visual emotional stimuli were set up. The collected stimuli allowed exploring the amount of emotional information

M.T. Riviello and A. Esposito, *On the Perception of Dynamic Emotional Expressions: A Cross-Cultural Comparison*, SpringerBriefs in Cognitive Computation 6, DOI 10.1007/978-94-024-0887-4_2

conveyed dynamically by the visual and auditory channels, and therefore identifying preferential channels exploited by humans in decoding emotional states as well as enlightening which are the emotional cues universally shared.

In a cross-cultural perspective, this work investigates if the ability to recognize emotional expressions as a function of the channel is also affected by the cultural context and in particular by the subject's native language. As already mentioned in Sect. 1.1.1, psychologists have long debated whether emotions are universal versus whether they vary across cultures. Our approach is based on the assumption that culture and language-specific paralinguistic patterns may influence the decoding process of emotional speech. In addition, the familiarity to the language and the subject's expositions to cultural norms and social rules may affect the recognition of emotional states in particular when they are vocally expressed.

To this aims, the emotional stimuli used were extracted from two different cultural context and played in two different languages: American English (as a global spread language) (Riviello et al. 2011) and Italian (as a country specific language). In addition, the participants involved in the experiments belong to 5 different, even though close, Western cultures, e.g. Italian, American, French, Hungarian and Lithuanian. In this context, two groups of participants were native speakers of the language and belong to the same cultural context of the administered stimuli. Their performance can be considered as a reference for an optimal identification of the emotional states under examination (Riviello and Esposito 2012).

2.1 Materials: The Cross-Modal Emotional Databases

The collected stimuli are based on extracts from American and Italian live recording movies (Esposito et al. 2009; Esposito and Riviello 2011), whose protagonists were carefully chosen among actors and actresses who are well regarded by critics and considered capable of giving some very real and careful interpretations. Differently from the other existing emotional databases proposed in literature, in this case, the actors/actresses had not been asked to produce an emotional expression for building an associated database, but rather they were acting according to a movie script, their performance was related and considered appropriate to a defined context as the movie director (supposed to be an expert) had judged. In addition, even though the emotions expressed in such video-clips were simulations under studio conditions (and may not have reproduced a genuine emotion but a stylized version of it) they were able to catch up and engage the emotional feeling of the spectators (the addressers) and therefore provided more confidence on the value of their perceptual emotional content. The stimuli were also noisy, as emotions experienced in real environments.

Each database consists of audio and video stimuli representing six emotional states: happiness, sarcasm/irony, fear, anger, surprise, and sadness. Except for sarcasm/irony, the remaining emotions are considered by many theories as primary ones and therefore universally shared (see Sect. 1.1).

For each database and for each of the emotional states under examination, 10 stimuli were identified, 5 expressed by an actor and 5 by an actress, for a total of 60 American and 60 Italian video-clips, each acted by a different actor and actress to avoid actor's bias in the ability to portray emotional states.

The stimuli were short (the average length was 3 s, SD = ± 1 s) to avoid the overlapping of emotional states that could confuse the subject's perception. Care was taken in selecting video clips where the protagonist's face and the upper part of the body were clearly visible. In addition, the semantic meaning of the produced utterances did not clearly express the portrayed emotional state and its intensity level was moderate. For example, the stimuli of sadness, where the actress/actor was clearly crying, or stimuli of happiness, where the protagonist was strongly laughing, were not included in the database. This was an attempt to allow the participants to observe less obvious emotional cues generally employed in a very natural and ecological setting, rather than in extreme emotional interactions.

The emotional labels assigned to the stimuli were first given by two experts and then by three naïve judges independently. The expert judges made a decision on the stimuli carefully exploiting emotional information within facial and vocal expressions such as a frame-by-frame analysis of changes in facial muscles, the rising and falling of F0 intonation contour, and the contextual situation the protagonist was interpreting. The naïve judges made their decision after watching the stimuli several times. There were no opinion exchanges between the experts and naïve judges and the final agreement on the labeling between the two groups was 100 %.

The collected stimuli extracted from movie scenes contain environmental noise and therefore are useful for testing realistic computer applications.

Both for the American and Italian data, the audio and mute video were extracted from each complete audio-video stimulus (video-clip) coming up with a total of 180 American and 180 Italian stimuli: 60 mute videos, 60 audio and 60 audio-video stimuli for each database.

2.2 Participants and Testing Procedure

The perceptual experiments involved 180 Italian, 180 American, 180 French, 180 Hungarian, and 180 Lithuanian participants. The participants' age was similar among countries, ranging from 18 to 35 years (26 ± 4.8).

Excluding the Americans, all participants have comparable knowledge of the English, since all of them used it as second language. The participants were volunteers principally recruited among university students.

For each group of 180, hence for each nationality, 90 participants were involved in the evaluation of the American database and 90 of the Italian database of stimuli. For each group of 90, 30 subjects evaluated the audio, 30 the mute video and 30 audio-video stimuli. Gender was equally balanced among the groups composed each by 15 males and 15 females.

The subjects were randomly assigned to the task and were required to carefully listen to and/or watch the stimuli via computer, wearing headphones, in a quiet room. They were instructed to pay attention to each presentation and decide which of the six emotional states were expressed.

Responses were recorded on a matrix paper form (60 × 8), where rows listed the stimuli numbers and columns the 6 selected emotional states (happiness, fear, anger, irony, surprise and sadness) plus the option for "others" indicating any other emotion not listed and the option for "no emotion", which was suggested when according to the subject's feeling the protagonist did not show emotions. The above-mentioned paper form was created in Italian and then translated in American, French, Hungarian and Lithuanian with the help of native speaker of the languages.

2.3 Data Analyses

The data obtained from participants of each nationality were first analyzed separately. For each database and for each set of stimuli (audio, mute video and audio-video) the frequencies of correct answers, intended here as the subjects' agreement on the label assigned to each stimulus, related to each emotion under consideration, were computed.

To assess the role of the perceptual mode affecting the identification of the emotional stimuli, repeated ANOVA measurements were performed on the frequencies of correct answers obtained by participants from each nationality, separately tested on the American and the Italian stimuli. To set up the analyses the *Perceptual mode* (audio, video, audio-video) was considered as a between subjects variable, while *Emotions* (happiness, fear, anger, irony, surprise and sadness) and *Actors' gender* (male, female) as within subjects variables. Significance was established for $\alpha = 0.05$.

Further analyses were performed separately on the data gathered from participants of each nationality to assess the role of the language and the cultural context characterizing the emotional expressions exploited as stimuli.

To this aim, for each perceptual mode (audio, video and audio-video) subjects' performance was assessed and compared on the set of American and Italian stimuli. In the repeated ANOVA measurements, the *Cultural context and the Language of the Stimuli* (American and Italian) was considered as between subjects variable, while *Emotions* and *Actors' gender* as within subjects variables. Significance was fixed for $\alpha = 0.05$.

To allow a cross-cultural comparison for establishing the effects of the cultural context and the language, further analyses were conducted. Six separated repeated ANOVA measurements for each perceptual mode were performed on the data obtained by each group of 30 American, Italian, French, Hungarian and Lithuanian subjects separately tested on the American and Italian audio, mute video and audio-video stimuli. In this case the ANOVAs' set up considered the subjects' *Nationality* as a between subjects variable and the *Emotions* and *Actor's Gender* as within subjects variables. Even in this case significance was established for $\alpha = 0.05$. All the statistical analyses were run using SPSS Statistic 17.0 software (SPSS 2008).

References

Esposito, A. (2007). The amount of information on emotional states conveyed by the verbal and nonverbal channels: Some perceptual data. In Y. Stylianou, M. Faundez-Zanuy, & A. Esposito (Eds.), *Progress in Nonlinear Speech Processing* (Vol. 4391, pp. 249–268), LNCS. Berlin Heidelberg: Springer. ISBN 978-3-540-71503-0.

Esposito, A. (2009). The perceptual and cognitive role of visual and auditory channels in conveying emotional information. *Cognitive Computation Journal, 1*, 268–278.

Esposito, A., & Riviello, M. T. (2011). The cross-modal and cross-cultural processing of affective information. In B. Apolloni, et al. (Eds.), *Frontiers in Artificial Intelligence and Applications* (Vol. 226, pp. 301–310). IOS press. ISBN: 978-1-60750-691-1 (print), ISBN 978-1-60750-692-8 (online).

Esposito, A., Riviello, M. T., & Di Maio, G. (2009). The COST 2102 Italian audio and video emotional database. In B. Apolloni, et al. (Eds.), *Frontiers in Artificial Intelligence and Applications* (Vol. 204, pp. 51–61). ISBN 978-1-60750-072-8.

Riviello, M. T., & Esposito, A. (2012). A cross-cultural study on the effectiveness of visual and vocal channels in transmitting dynamic emotional information. *Acta Polytechnica Hungarica, Journal of Applied Sciences, 9*(1), 157–170. ISSN 1785-8860.

Riviello, M. T., Chetouani, M., Cohen, D., & Esposito, A. (2011). On the perception of emotional "voices": A cross-cultural comparison among American, French and Italian subjects. In A. Esposito, et al. (Eds.), *Analysis of Verbal and Nonverbal Communication and Enactment: The Processing Issue* (Vol. 6800, 368–377), LNCS. Springer. ISBN 978-3-642-25774-2.

SPSS Inc. Released. (2008). SPSS Statistics for Windows, Version 17.0. Chicago: SPSS Inc.

Chapter 3
Results for Italian Participants

Abstract This chapter reports the results of perceptual experiments involving Italian subjects in the assessment of emotional stimuli extracted from Italian (as a country-specific language) and American English (as a global spread language) live recording movie scenes. The stimuli concern some of the basic emotions of happiness, fear, anger, surprise, sadness, as well as a language specific emotion such as sarcasm/irony. In order to investigate the effects of the communication channels on the emotional decoding process, the stimuli are portrayed through the visual, auditory, and the audiovisual mode. The main goal was to investigate whether, for Italian subjects, the visual channel is more effective than the auditory one to infer emotional information and if this effectiveness is affected by the cultural context and in particular by the language. Results show that the audio and visual components of emotional messages convey much the same amount of information either separately or in combination, suggesting that emotional information is not added over the amount of cues provided. In addition, Italian subjects favor vocal information while decoding vocal emotional information in their native language and visual information in the non-native cultural context, supporting the hypothesis that emotional communication is affected by culture and language. Finally, among the basic emotions considered, anger got the higher decoding accuracy.

Keywords Emotional recognition · Dynamic expressions · Cultural effects · Emotion categories · Native italian speakers · Cross-Cultural experiments · Language familiarity · Cross-modal database · Auditory emotional information · Visual emotional information

3.1 Recognition Performance on the American and Italian Cross-Modal Databases

Figures 3.1 and 3.2 report the results obtained by each group of 30 Italian subjects, tested separately on the American (Fig. 3.1) and Italian (Fig. 3.2) audio, video and audio-video emotional stimuli. On the x-axis are the basic emotions under

M.T. Riviello and A. Esposito, *On the Perception of Dynamic Emotional Expressions: A Cross-Cultural Comparison*, SpringerBriefs in Cognitive Computation 6, DOI 10.1007/978-94-024-0887-4_3

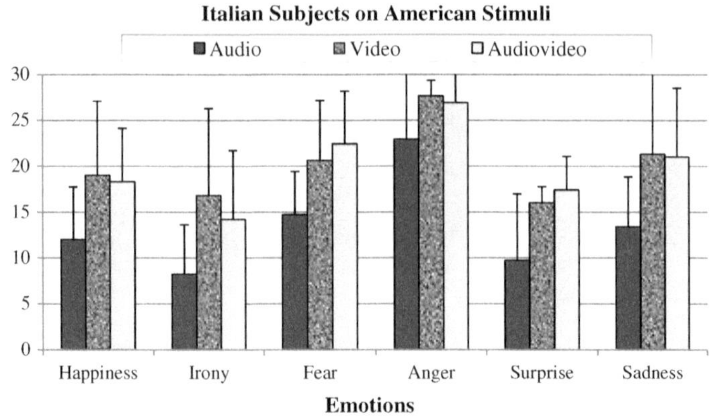

Fig. 3.1 Mean number of correct responses (and standard deviations) to the American emotional stimuli portrayed through the audio, mute video, and combined audio/video to 3 groups each composed of 30 Italian subjects. The data are reported for each emotion and the three experimental conditions

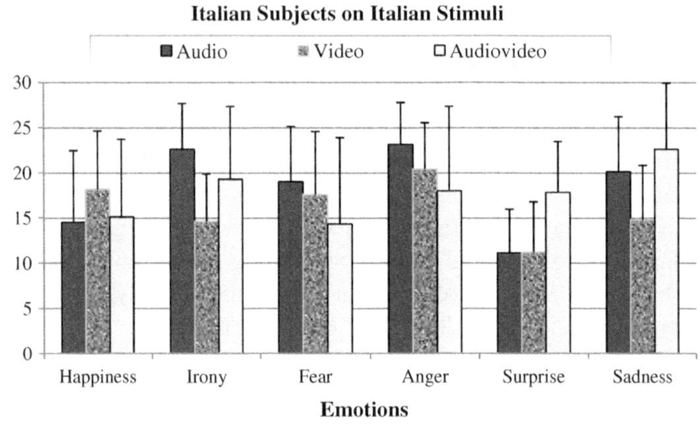

Fig. 3.2 Mean number of correct responses (and standard deviations) to the Italian emotional stimuli portrayed through the audio, mute video, and combined audio/video to 3 groups each composed of 30 Italian subjects. The data are reported for each emotion and the three experimental conditions

consideration and on the y-axis for each emotion it is reported the mean number of correct responses and the associated standard deviations under the three experimental conditions.

Repeated ANOVA (SPSS 2008) measurements were performed on the data obtained by the Italian participants tested on the two databases of emotional stimuli, separately. In the analyses, the *Perceptual mode* (audio, video, audio-video) was considered as a between subjects variable, while *Emotions* (happiness, fear, anger, irony, surprise and sadness) and *Actors' gender* (male, female) as within subjects variables. Significance was established for $\alpha = 0.05$.

The ANOVA show that the *Perceptual mode* plays a significant role for the evaluation of both American (F (2, 12) = 25.375, ρ = 0.000), and the Italian (F (2, 12) = 5.884, ρ = 0.017) stimuli.

For the American stimuli, the mute video and audio-video communication modes convey the same amount of emotional information (F (1, 8) = 0.025, ρ = 0.87), while significant differences were found between the audio and audio-video (F (1, 8) = 53.075, ρ = 0.0001) and the audio and mute video (F (1, 8) = 33.722, ρ = 0.000).

For the Italian stimuli, significant differences were found between the audio and mute video (F (1, 8) = 8.436, ρ = 0.020), and the mute video and the audio-video (F (1, 8) = 77.58 ρ = 0.024) modalities.

Differences are not significant, between the audio and the combined audio- video (F (1, 8) = 0.678, ρ = 0.434) modes.

The identification of both American (F (5, 60) = 17.472, ρ = 0.0001) and Italian (F (5, 60) = 3.675, p = 0.006) stimuli is affected by the specific portrayed emotion.

Anger is the emotion with the highest percentage of correct recognition, for both the American (m: 25.8, SD: 3.9) and Italian (m: 20.5, SD: 6.37) emotional stimuli. Irony is hardly recognized through the American emotional expressions (m: 13.067, SD: 6.5), as well as surprise both trough the American (m: 14.37, SD: 5.3) and Italian (m: 13.367, SD: 4.3) ones.

The biggest difference in the recognition accuracy is for irony, that is very well identified in the native cultural context mostly in the audio (for Italian audio stimuli: m: 22.6, SD: 5.2; for American audio stimuli: m: 8.2, SD: 4.9) probably because the expressions of this emotion are strictly linked to the cultural context.

When evaluating the Italian stimuli, where the audio modality is preferential to infer emotional data, the emotional cues of happiness is be better decoded through the visual mode (audio m: 14.5, SD: 6.5; video m: 18.2 SD: 5.6; audio-video: m: 15.1, SD: 9.06). Emotion recognition accuracy is not affected by the gender of the protagonist portraying the emotional expressions both for the American (F (1, 12) = 0.906, ρ = 0.360) and Italian F (1, 12) = 1.079, ρ = 0.319) data.

3.2 Assessment of the Cultural Effects

Figure 3.3 groups together the Italian subjects (divided in two groups each of 30 subjects) number of correct responses obtained separately on the American and Italian audio, mute video and audio-video respectively. It clearly appears that the visual channel is preferred when subjects are tested on the non-native language and cultural context.

ANOVA analyses were performed on the above data, considering the *Cultural context and the Language of the Stimuli* (American and Italian) as between

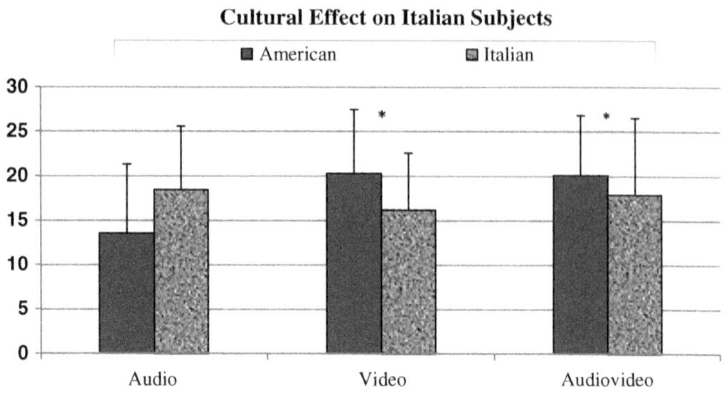

Fig. 3.3 Mean number of correct responses (and standard deviations) obtained by each 2 groups of 30 Italian subjects tested on the American and Italian audio, video, and audio-video emotional stimuli, separately

subjects variable, while *Emotions* and *Actors' gender* as within subjects variables. Significance was established for $\alpha = 0.05$.

The analyses results show that for Italian subjects do exist a difference in perceiving emotional information when it is expressed through Italian with respect to American voices (F (1, 8) = 32.178, $\rho = 0.000$). As it should be expected, Italians had better identify emotions through the Italian (m: 18.4, SD: 5.5) rather than the American (m: 13.48, SD: 6.3) audio emotional stimuli.

Differences were also found in the perception of emotional information through the American and Italian mute video (F (1, 8) = 13.87, $\rho = 0.006$) and audio-video (F (1, 8) = 9.47, $\rho = 0.015$) stimuli. Nevertheless, Italian subjects seem to better identify emotions through the American (mute video: m: 20.28, SD: 5.4, and audio-video stimuli: m: 20.03, SD: 5.2) rather than the Italian visual information (mute video: m: 16.133, SD: 5.8; and audio-video: m: 17.85, SD: 8.4).

3.3 Comments

The results show that Italian subjects are able to recognize the proposed emotions in both the American and the Italian cultural and linguistic context. However, they favored audio information when they are native speakers of the language (Esposito et al. 2009) and video and audio-video information in the non-native cultural context, being able to obtain the same performance even through different communication modes.

The data also reveal that the audio and visual components of emotional messages convey much the same amount of information either separately or in combination, hence suggesting that each channel performs a robust encoding of the emotional features.

The results of the comparison on American and Italian audio, mute video and audio-video stimuli confirmed the remarkable role of the Italian vocal expressions in conveying emotions for the Italian participants, but also reveal the strong efficacy of American visual expressions in transmitting emotional information: the Italian subjects perform significantly better on the American than on the Italian visual emotional stimuli.

References

Esposito, A., Riviello, M. T., Bourbakis, N. (2009). Cultural specific effects on the recognition of basic emotions: A study on Italian subjects. In A. Holzinger, K. Miesenberger (Eds.), *USAB 2009* (Vol. 5889, pp. 135–148), LNCS. Berlin Heidelberg: Springer. ISBN: 978-3-642-10307-0.

SPSS Inc. Released. (2008). SPSS Statistics for Windows, Version 17.0. Chicago: SPSS Inc.

Chapter 4
Results for American Participants

Abstract This chapter reports the results of perceptual emotional experiments involving American subjects in the assessment of emotional stimuli extracted from Italian (as a country-specific language) and American English (as a global spread language) live recording movie scenes. The stimuli concern some of the basic emotions of happiness, fear, anger, surprise, sadness, as well as a language specific emotion such as sarcasm/irony, and are portrayed through the mute video, the audio alone, and the combined audio/video mode. The main goal was to investigate whether, for American subjects, the visual channel is more effective than the auditory one to infer emotional information and whether this effectiveness is affected by the cultural context and in particular by the language. Results show that American subjects are facilitated by the visual mode in the identification of emotional information, independently from the cultural context and language of the stimuli. Furthermore, the bimodal presentation (the audio/video combined modality) of emotional information did not improve significantly the recognition accuracy with respect to the mute video modality. The familiarity with the Italian culture and language seem to play a role in the recognition performance. Among the basic emotions considered, anger is perceptually privileged because received the highest percentage of recognition accuracy.

Keywords Emotional recognition · Dynamic expressions · Cultural effects · Emotion categories · Native american speakers · Cross-Cultural experiments · Language familiarity · Cross-modal database · Auditory emotional information · Visual emotional information

4.1 Recognition Performance on the American and Italian Cross-Modal Databases

Figures 4.1 and 4.2 report the emotion recognition accuracy obtained by the different groups of American subjects tested separately in identifying the six proposed emotions through the American (Fig. 4.1) and Italian (Fig. 4.2) audio, mute video, and audio-video stimuli.

M.T. Riviello and A. Esposito, *On the Perception of Dynamic Emotional Expressions: A Cross-Cultural Comparison*, SpringerBriefs in Cognitive Computation 6, DOI 10.1007/978-94-024-0887-4_4

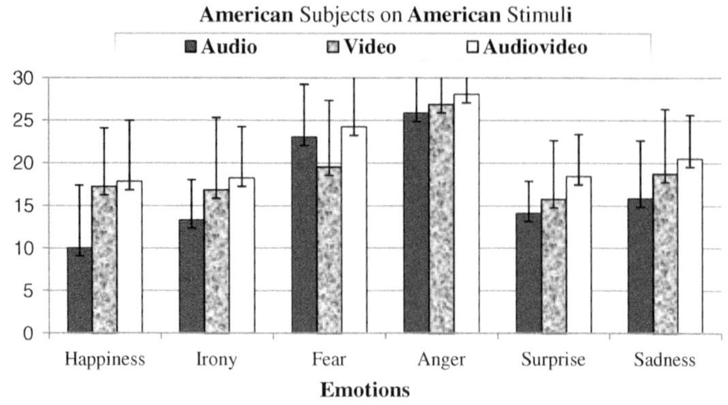

Fig. 4.1 Mean number of correct responses (and standard deviations) to the American emotional stimuli portrayed through the audio, mute video, and combined audio/video to 3 groups each composed of 30 American subjects. The data are reported for each emotion and the three experimental conditions

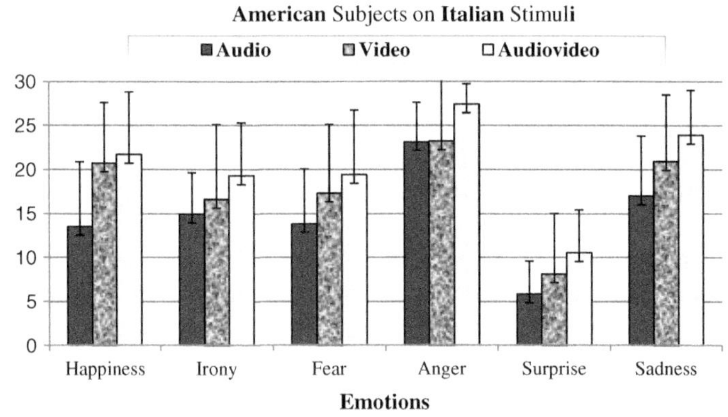

Fig. 4.2 Mean number of correct responses (and standard deviations) to the Italian emotional stimuli portrayed through the audio, mute video, and combined audio/video to 3 groups each composed of 30 American subjects. The data are reported for each emotion and the three experimental conditions

The ANOVA analyses (SPSS 2008) (as described in Sect. 3.1) executed on the data obtained by the American participants evaluating the two databases of emotional stimuli reveal that the *Perceptual mode* plays a significant role for the evaluation of both the American ($F(2, 12) = 4.090$, $\rho = 0.044$), and Italian emotional stimuli ($F(2, 12) = 7.241$, $\rho = 0.009$). Significant differences, were found between the audio and audio-video communication modalities both for the American ($F(1, 8) = 9.031$, $\rho = 0.017$), and Italian ($F(1, 8) = 21.674$, $\rho = 0.002$) stimuli. The American and Italian audio and mute video ($F(1, 8) = 1.696$, $\rho = 0.229$ for American, and $F(1, 8) = 4.114$, $\rho = 0.077$ for Italian stimuli), and the American and Italian mute video and audio-video (F (1, 8) = 2.323, ρ = 0.166, for American, and F (1, 8) = 2.303, ρ = 0.168 for Italian stimuli) convey the same amount of emotional information independently from stimuli's cultural context and language.

The identification of both American (F (5, 60) = 16.217, ρ = 0.000) and Italian (F (5, 60) = 28.779, ρ = 0.000) emotional stimuli is affected by the specific portrayed emotion.

Anger is the emotion with the highest percentage of correct recognition, both for the American (m: 26.67, SD: 3.15) and Italian (m: 24.567, SD: 5.06) vocal and visual emotional stimuli. Generally, surprise is not easy to recognize, in particular through the Italian (m: 8.133, SD: 5.3) stimuli, independently from the perceptual modality.

Fear, when portrayed through the American expressions, seems to be very well identified through the audio and the audio-video, rather than the mute video (audio: m: 23, SD: 6.06; video: m 19.5, SD: 7.8, audio-video: m: 24.2, SD: 5.2, stimuli).

In addition, happiness seems to be more easily identified through the visual mode both for the American (audio: m: 10, SD: 7.18; video: m: 17.2, SD: 7.2, audio-video: 17.8, SD: 9), and the Italian (audio: m: 13.5, SD: 7.37; video: m: 20.7, SD: 4.87, audio-video: 21.7, SD: 5.9) emotional stimuli. The actors' gender does not affect the recognition accuracy for the American (F (1, 12) = 0.983, ρ = 0.34) and the Italian (F (1, 12) = 3.571, ρ = 0.083) data.

4.2 Assessment of the Cultural Effects

The American subjects' performance on the American and Italian audio, mute video and combined audio-video is reported in Fig. 4.3.

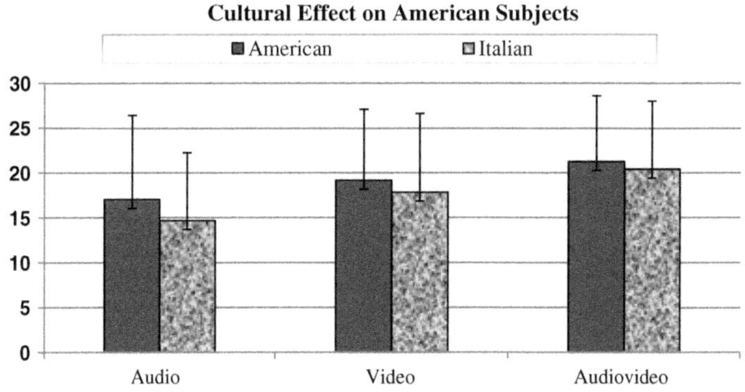

Fig. 4.3 Mean number of correct responses (and standard deviations) obtained by each 2 groups of 30 American subjects tested on the American and Italian audio, video, and audio-video emotional stimuli, separately

The ANOVA (as described in Sect. 3.2) results reveal that for American subjects there were no differences in their ability in identifying emotional expressions through the American and Italian audio ($F\ (1,\ 8) = 2.87$, $\rho = 0.128$), mute video ($F\ (1,\ 8) = 0.553$, $\rho = 0.479$), and audio-video ($F\ (1,\ 8) = 0.427$, $\rho = 0.532$) stimuli.

4.3 Comments

In contrast with what it was expected, i.e. that American subjects would have better performance in identifying emotional information from their native language, and therefore in the audio alone as it has been shown for Italian subjects, these results reveal that American subjects are more accurate in identifying emotional information through the visual mode, independently from the cultural context and language of the stimuli (Esposito and Riviello 2011).

The results also show that adding emotional information (i.e. considering the American subject performance on the combined auditory and visual channel) did not improve significantly the recognition accuracy with respect the single mute video modality.

In addition, American subjects seem to identify equally emotional information both through the American and Italian audio, mute video and audio-video stimuli. These results could be due to the close relationship and continuous exchanges between the American and Italian cultures.

References

Esposito, A., & Riviello, M. T. (2011). The cross-modal and cross-cultural processing of affective information. In B. Apolloni, et al. (Eds.), *Frontiers in Artificial Intelligence and Applications* (Vol. 226, pp. 301–310). IOS press. ISBN: 978-1-60750-691-1 (print), ISBN 978-1-60750-692-8 (online).

SPSS Inc. Released. (2008). SPSS Statistics for Windows, Version 17.0. Chicago: SPSS Inc.

Chapter 5
Results for French Participants

Abstract This chapter reports the results of perceptual experiments that involved French subjects assessing emotional stimuli extracted from Italian (as a country—specific language) and American English (as a global spread language) live recording movie scenes. The stimuli concern some of the basic emotions of happiness, fear, anger, surprise, sadness, as well as a language specific emotion such as sarcasm/irony, and are portrayed through the mute video, the audio alone, and the combined audio/video mode. The main goal was to investigate whether the visual channel is more effective than the auditory one to infer emotional information and whether this effectiveness is affected by the cultural context and in particular by the language. The results reveal that French subjects better identify emotions through visual than through vocal information, the recognition accuracy does not increase when bimodal modality is exploited and language affects the recognition performance. In addition, among the basic emotions considered, anger is perceptually privileged because received the highest percentage of recognition accuracy.

Keywords Emotional recognition · Dynamic expressions · Cultural effects · Emotion categories · Cross-Cultural experiments · Language familiarity · Cross-modal database · Emotional database · Auditory emotional information · Visual emotional information

5.1 Recognition Performance on the American and Italian Cross-Modal Databases

The results for the French participants are displayed in Figs. 5.1 and 5.2 that respectively report the French recognition accuracy on the American and Italian emotional expressions portrayed through the audio, mute video and audio-video modes.

The ANOVA (SPSS 2008) (as described in Sect. 3.1) results for French participants tested on the two databases of emotional stimuli show that the identification of emotional expressions is affected by the *Perceptual mode* both for

M.T. Riviello and A. Esposito, *On the Perception of Dynamic Emotional Expressions: A Cross-Cultural Comparison*, SpringerBriefs in Cognitive Computation 6, DOI 10.1007/978-94-024-0887-4_5

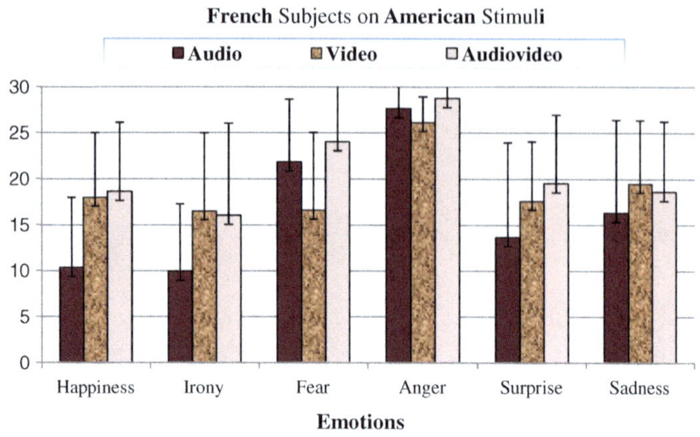

Fig. 5.1 Mean number of correct responses (and standard deviations) to the American emotional stimuli portrayed through the audio, mute video, and combined audio/video to 3 groups each composed of 30 French subjects. The data are reported for each emotion and the three experimental conditions

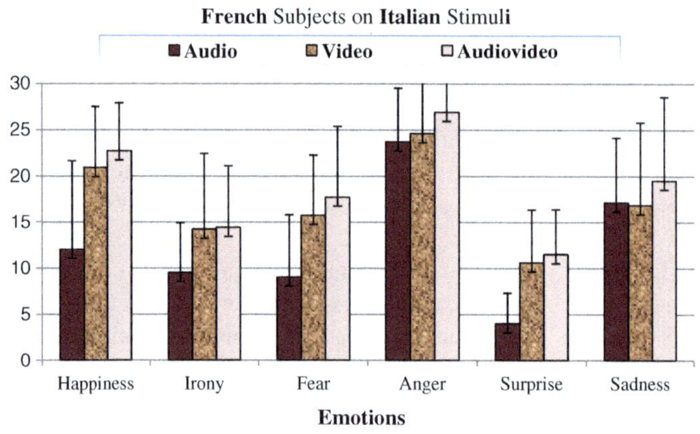

Fig. 5.2 Mean number of correct responses (and standard deviations) to the Italian emotional stimuli portrayed through the audio, mute video, and combined audio/video to 3 groups each composed of 30 French subjects. The data are reported for each emotion and the three experimental conditions

American (F (2, 12) = 4.397, ρ = 0.037), and the Italian stimuli (F (2, 12) = 8.339, ρ = 0.005).

No significant differences were found between the mute video and the audio-video modalities for both the American (F (1, 8) = 1.426, ρ = 0.267), and the Italian (F (1, 8) = 0.771, ρ = 0.406) emotional expressions, while significant differences do exist between the audio and audio-video modalities for both the American (F (1, 8) = 8.493, ρ = 0.019) and Italian (F (1, 8) = 28.793, ρ = 0.001) emotional stimuli.

The audio and mute video communication modes convey the same amount of emotional information when American stimuli are exploited ($F(1, 8) = 3.370$, $\rho = 0.104$), whereas significant differences were found between the two modalities in conveying Italian emotional information ($F(1, 8) = 8.004$, $\rho = 0.022$).

The specific portrayed emotion affects the identification of both the American ($F(5, 60) = 15.151$, $\rho = 0.037$) and Italian ($F(5, 60) = 23.096$, $\rho = 0.000$) emotional stimuli.

Anger is the emotion with the highest correct recognition, both for the American (m: 27.467, SD: 5.7) and Italian (m: 25.067, SD: 5.4) stimuli. Irony (m: 14.133, SD: 7.1), happiness (m: 15.63, SD: 7.18), and surprise (m: 16.9, SD: 7.6) are scarcely recognized through the American emotional expressions, independently from the conveying channel. This is also true for surprise (m: 8.7, SD: 4.7) and irony (m: 12.7, SD: 6.6) expressed through Italian emotional stimuli.

The American emotional cues of fear, seem to be better identified through the audio and audio-video modality (audio: m: 21.8, SD: 7.2; video: m: 16.6, SD: 8.8; audio-video: m: 24, SD: 6).

Furthermore, for French participants happiness is better identified in the mute video than in the audio and in the audio-video for both the American and Italian set of emotional stimuli (American audio: m: 10.3, SD: 6.6; video: m: 18, SD: 7.2, audio-video: 18.6, SD: 7.7, and Italian audio: m: 12, SD: 9.7; video: m: 20.9, SD: 5.1, audio-video: 22.7, SD: 4.2).

For the American stimuli, the gender of the protagonist expressing the proposed emotions does not affect the identification of the emotional information ($F(1, 12) = 1.158$, $\rho = 0.303$), whereas it slightly influences the recognition of the Italian data ($F(1, 12) = 4.816$, $\rho = 0.049$).

5.2 Assessment of the Cultural Effects

Figure 5.3 reports the results obtained by comparing each two groups of 30 French subjects tested on American and Italian audio, video and audio-video.

The ANOVA (as described in Sect. 3.2) results from French subjects compared on their ability in perceiving emotions through the American and Italian stimuli, show that a difference do exist for the audio modality ($F(1, 8) = 15.94$, $\rho = 0.004$). The American vocal emotional expressions (m: 16.58, SD: 7.3) are more accurately recognized then the Italian ones (m: 12.55, SD: 6.27). No differences were found between the American and Italian mute video ($F(1, 8) = 1.078$, $\rho = 0.33$) and audio-video ($F(1, 8) = 1.776$, $\rho = 0.219$) in transmitting emotional information.

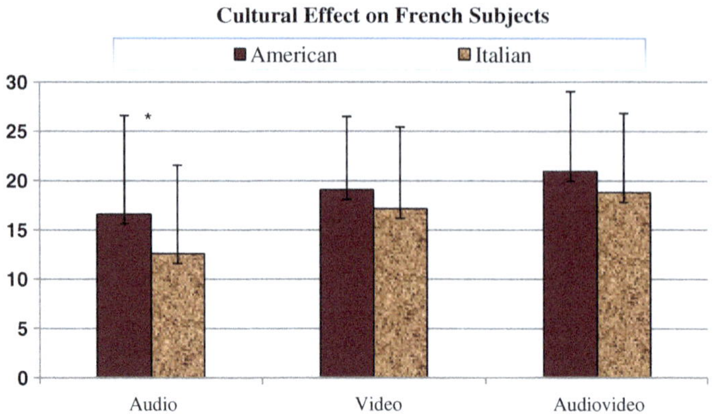

Fig. 5.3 Mean number of correct responses (and standard deviations) obtained by each 2 groups of 30 French subjects tested on the American and Italian audio, video, and audio-video emotional stimuli, separately

5.3 Comments

The data obtained by French participants confirmed that emotional information is not added over the amount of emotional signals provided. The audio-video modality conveys the same amount of emotional information of the mute video. As for American, French subjects prefer are more accurate in inferring emotional information both through American and Italian visual stimuli (Riviello et al. 2011).

In addition, French subjects do better extract emotional information from American than from Italian vocal expressions, whereas no differences do exist in their ability to decode visual emotional information when both American and Italian video and audio-video stimuli are evaluated.

References

Riviello, M. T., Chetouani, M., Cohen, D., & Esposito, A. (2011). Inferring emotional information from vocal and visual cues: A cross-cultural comparison. Best paper award. In *IEEE 2nd International Conference on Cognitive Computation, Budapest* (pp. 1–4). E-ISBN: 978-963-8111-78-4, Print ISBN: 978-1-4577-1806-9.

SPSS Inc. (2008). SPSS Statistics for Windows, Version 17.0. Chicago: SPSS Inc.

Chapter 6
Results for Hungarian Participants

Abstract This chapter reports the results of perceptual experiments that involved Hungarian subjects assessing emotional stimuli extracted from Italian (a country-specific language) and American English (as a global spread language) live recording movie scenes. The stimuli concern some of the basic emotions of happiness, fear, anger, surprise, sadness, as well as a language specific emotion such as sarcasm/irony, are portrayed through the mute video, audio alone, and combined audio/video mode. The main goal was to investigate whether the visual channel is more effective than the auditory one to infer emotional information and whether this effectiveness is affected by the cultural context and in particular by the language. Results show that Hungarian subjects are more accurate in inferring emotional information from visual rather than from vocal cues. Language plays a role in the recognition task since they perform better on the American rather than Italian vocal expressions. In addition, the audiovisual presentation of the emotional did not significantly improve subjects' recognition accuracy. Finally, among the basic emotions anger seems to be perceptually privileged.

Keywords Emotional recognition · Dynamic expressions · Cultural effects · Emotion categories · Decoding processing · Cross-Cultural experiments · Language familiarity · Cross-modal emotional database · Auditory emotional information · Visual emotional information

6.1 Recognition Performance on the American and Italian Cross-Modal Databases

Figures 6.1 and 6.2 show the agreement obtained by the Hungarian subjects evaluating emotional expressions through American and Italian cross-modal emotional stimuli.

The ANOVA analyses (SPSS 2008) (as described in Sect. 3.1) for the data obtained by Hungarian participants at the perceptual experiments show that the *perceptual mode* affects the recognition of both the American ($F\ (2, 12) = 10.455$,

M.T. Riviello and A. Esposito, *On the Perception of Dynamic Emotional Expressions: A Cross-Cultural Comparison*, SpringerBriefs in Cognitive Computation 6, DOI 10.1007/978-94-024-0887-4_6

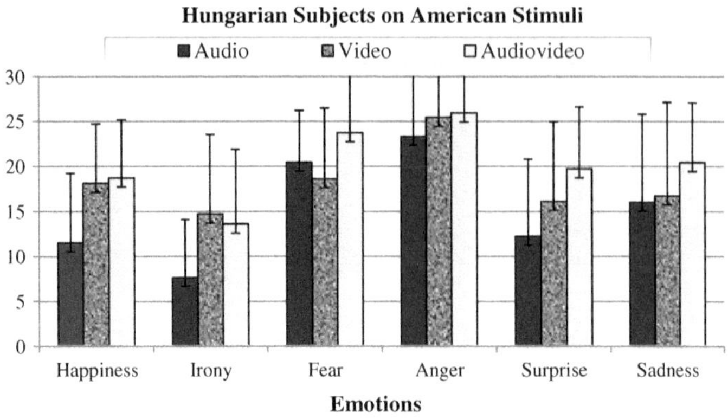

Fig. 6.1 Mean number of correct responses (and standard deviations) to the American emotional stimuli portrayed through the audio, mute video, and combined audio/video to 3 groups each composed of 30 Hungarian subjects. The data are reported for each emotion and the three experimental conditions

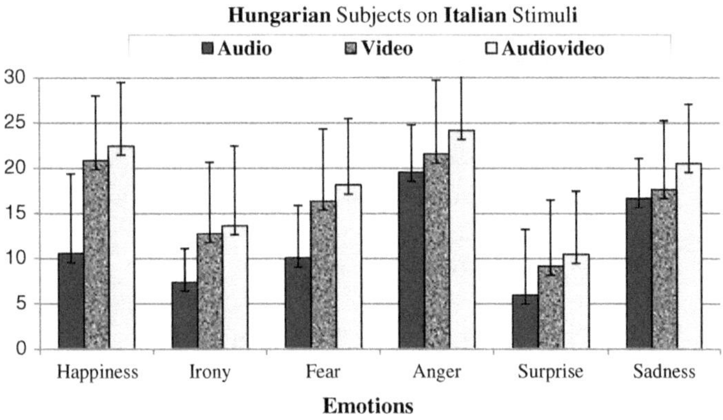

Fig. 6.2 Mean number of correct responses (and standard deviations) to the Italian emotional stimuli portrayed through the audio, mute video, and combined audio/video to 3 groups each composed of 30 Hungarian subjects. The data are reported for each emotion and the three experimental conditions

$\rho = 0.002$), and Italian ($F(2, 12) = 9.223$, $\rho = 0.004$) emotional expressions. The mute video and the audio-video communication modes convey the same amount of emotional information for the American ($F(1, 8) = 2.749$, $\rho = 0.136$) as well as for the Italian ($F(1, 8) = 0.996$, $\rho = 0.347$) emotional stimuli.

Slightly significant differences were found between the American and Italian audio and mute video (American stimuli: $F(1, 8) = 6.789$, $\rho = 0.031$; Italian stimuli: $F(1, 8) = 10.145$, $\rho = 0.013$), and the American and Italian audio and audio-video (American stimuli: $F(1, 8) = 29{,}276$, $\rho = 0.001$; Italian stimuli: $F(1, 8) = 23{,}761$, $\rho = 0.001$).

The recognition of a given emotional states significantly depends on the portrayed emotion (American stimuli: F (5, 60) = 11.462, ρ = 0.000; Italian stimuli: F (5, 60) = 20.433, ρ = 0.000).

Independently from the perceptual mode, anger is the perceptually privileged emotion, both in the American (24.867 SD: 4.7) and Italian (m: 20.5, SD: 6.7) emotional expressions, while irony and surprise are not easily recognized both through the American (irony: m: 11.96, SD: 6.7; surprise: m: 16, SD: 8.2), and Italian (irony: m: 11.2, SD: 7; surprise: m: 8.467, SD: 7) vocal and visual emotional stimuli. Emotional information of fear are slightly better identified through vocal cues when the American stimuli are exploited (audio: m: 20.4; SD: 5; video: m: 18.6, SD: 8.2; audio-video: m: 23.7; SD: 7), while happiness is again better expressed by visual information (American audio: m: 11.5, SD: 7.3; video: m: 18.1, SD: 6.5, audio-video: 6.12, SD: 1.628, and Italian audio: m: 10.5, SD: 7.7 video: m: 20.8, SD: 4.7, audio-video: 22.4, SD: 5.05).

Emotion recognition accuracy is not affected by the gender of the protagonist portraying the American emotional expressions (F (1, 12) = 0.349, ρ = 0.566), this is not true for the Italian emotional stimuli (F (1, 12) = 7.588, ρ = 0.017).

6.2 Assessment of the Cultural Effects

The Hungarian subjects' results comparing the recognition scores on the American and Italian audio, mute video and combined audio-video are reported in Fig. 6.3.

The ANOVA analyses (as described in Sect. 3.2) reveal a significant difference between the Hungarian recognition accuracy for the American and Italian audio stimuli (F (1, 8) = 18.99, ρ = 0.002): American vocal emotional expressions (m: 15.17, SD: 7.1) are better identified than the Italian ones (m: 11.63, SD: 5.4).

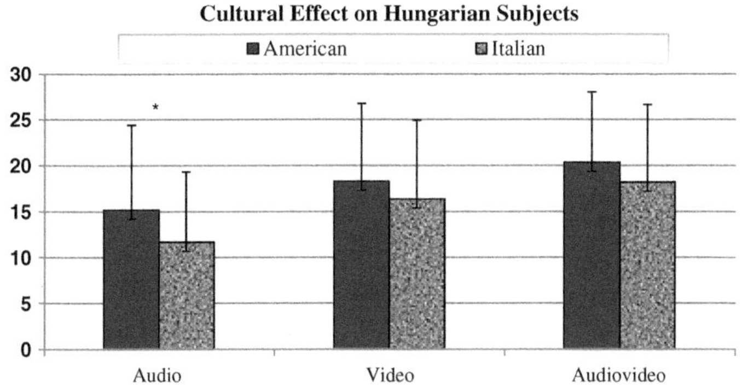

Fig. 6.3 Mean number of correct responses (and standard deviations) obtained by each 2 groups of 30 Hungarian subjects tested son the American and Italian audio, video, and audio-video emotional stimuli, separately

No significant differences were found between the American and Italian mute video ($F(1, 8) = 1.175$, $\rho = 0.292$) and audio-video ($F(1, 8) = 2.248$, $\rho = 0.172$) emotional stimuli in transmitting emotional information.

6.3 Comments

Data show that Hungarian subjects had better decode emotional information from visual rather than from vocal cues. This preference is stronger when Hungarian subjects are tested on Italian emotional stimuli rather than on American ones Riviello et al. (2012).

The bimodal presentation of the emotional information did not improve subjects' recognition accuracy: there were no differences between the mute video and the audio-video modalities in conveying emotions for both American and Italian stimuli. Hungarian subjects perform better on the American rather than Italian vocal expressions, while showing similar recognition accuracy for both visual American and Italian data.

References

Riviello, M. T., Esposito, A., & Vicsi, K. (2012). A cross-cultural study on the perception of emotions: How Hungarian subjects evaluate American and Italian emotional expressions. In A. Esposito, A. M. Esposito, A. Vinciarelli, R. Hoffman, & V. C. Müller (Eds.) *"Behavioural cognitive systems", cognitive behavioural systems, LNCS* 7403, 424–433. Berlin Heidelberg: Springer. ISBN: 978-3-642-34583-8. doi:10.1007/978-3-642-34584-5_38.

SPSS Inc. Released (2008). SPSS statistics for windows, version 17.0. Chicago: SPSS Inc.

Chapter 7
Results for Lithuanian Participants

Abstract This chapter reports the results of perceptual experiments that involved Lithuanian subjects assessing emotional expressions extracted from Italian (a country-specific language) and American English (as a global spread language) live recording movie scenes. The stimuli concern some of the basic emotions of happiness, fear, anger, surprise, sadness, as well as a language specific emotion such as sarcasm/irony, and are portrayed through the mute video, the audio alone, and the audiovisual mode. The main goal was to investigate whether the visual channel is more effective than the auditory one to infer emotional information and whether this effectiveness is affected by the cultural context and in particular by the language. Results show that visual signals allow a more effective identification of emotions then vocal ones no matter the American and Italian nature of the emotional expressions. Also, bimodality (audio/video combined mode) does not improve the Lithuanian subjects ability in inferring emotional information and a language affects is displayed since the American emotional vocal expressions are more accurately recognized than the Italian ones.

Keywords Emotional recognition · Dynamic expressions · Cultural effects · Emotion categories · Cross-Cultural experiments · Language familiarity · Cross-modal database · Emotional database · Vocal emotional information · Visual emotional information

7.1 Recognition Performance on the American and Italian Cross-Modal Databases

The results for Lithuanian participants are displayed in Figs. 7.1 and 7.2, respectively reporting their recognition accuracy on the American and Italian emotional expressions portrayed through the audio mute video and audio-video modes.

The ANOVA (SPSS 2008) (as described in Sect. 3.1) performed on the data obtained by the Hungarian subjects assessing show that the *perceptual mode* makes a difference in the recognition of emotional information both for the American

M.T. Riviello and A. Esposito, *On the Perception of Dynamic Emotional Expressions: A Cross-Cultural Comparison*, SpringerBriefs in Cognitive Computation 6, DOI 10.1007/978-94-024-0887-4_7

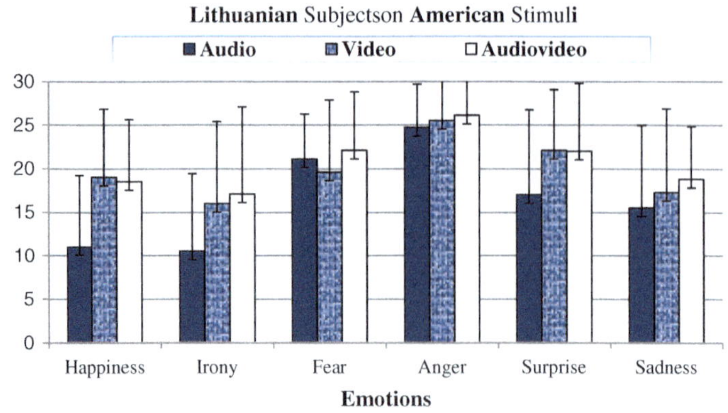

Fig. 7.1 Mean number of correct responses (and standard deviations) to the American emotional stimuli portrayed through the audio, mute video, and combined audio/video to 3 groups each composed of 30 Lithuanian subjects. The data are reported for each emotion and the three experimental conditions

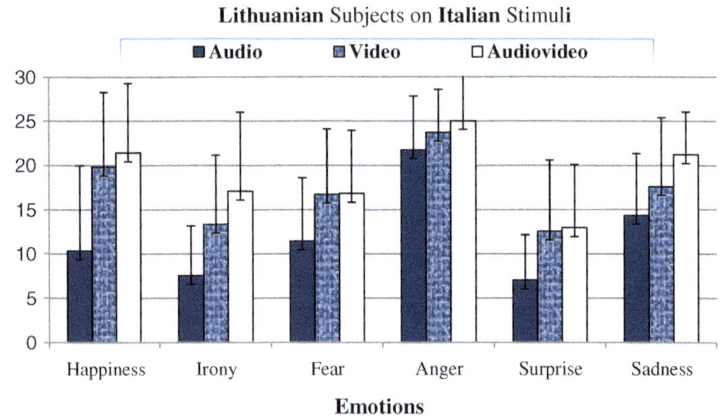

Fig. 7.2 Mean number of correct responses (and standard deviations) to the Italian emotional stimuli portrayed through the audio, mute video, and combined audio/video to 3 groups each composed of 30 Lithuanian subjects. The data are reported for each emotion and the three experimental conditions

(F (2, 12) = 5.673, ρ = 0.018), and the Italian (F (2, 12) = 9.549, ρ = 0.003) emotional stimuli.

No significant differences, in transferring emotional information, were found between the mute video and audio-video modalities both for the American (F (1, 8) = 0.421, ρ = 0.535) and Italian (F (1, 8) = 0.851, ρ = 0.383) emotional data.

Significant differences were found between the audio and audio-video modes for the American (F (1, 8) = 12.976, ρ = 0.007), and Italian stimuli (F (1, 8) = 27.525, ρ = 0.001), as well as between the audio and mute video channel for the American ($F(1, 8) = 5.372$, $ρ = 0.049$) and the Italian stimuli ($F(1, 8) = 9.835$, $ρ = 0.014$).

The specific portrayed emotion affects the identification of both the American ($F(5, 60) = 11.462$, $ρ = 0.000$) and Italian ($F(5, 60) = 14.775$, $ρ = 0.000$) emotional stimuli.

Among the emotional states proposed anger is the emotion with the highest percentage of correct recognition, both in the American (m: 25,433, SD: 4.4) and Italian (m: 23.467, SD: 5.6) total data. Irony get the lowest accuracy rate both for the American (m: 14.53; SD: 7.4), and Italian (m: 12.6; SD: 7.8) visual and vocal emotional expressions, just like surprise trough Italian (m: 10.8; SD; 6.8) stimuli.

Emotional information of fear are slightly better identified through vocal cues when the American stimuli are exploited (audio: m: 21.1; SD: 5; video: m: 8.6, SD: 2.412; audio-video: m: 22.1; SD: 6.6), while happiness is better expressed by visual information (American audio: m: 11, SD: 7.3; video: m: 19, SD: 8, audio-video: 18.5, SD: 7.3, and Italian audio: m: 10.3, SD: 9.4; video: m: 19.8, SD: 6.3, audio-video: 21.4, SD: 5.6).

The gender of the protagonist portraying the emotional expressions does not affect the identification of the emotional information of the American stimuli ($F(1, 12) = 2.202$, $ρ = 0.164$), whereas it plays a role in the recognition of the Italian data ($F(1, 12) = 13.686$, $ρ = 0.003$).

7.2 Assessment of the Cultural Effects

Figure 7.3 reports the results obtained comparing Lithuanian subjects tested on American and Italian audio, mute video and audio-video stimuli.

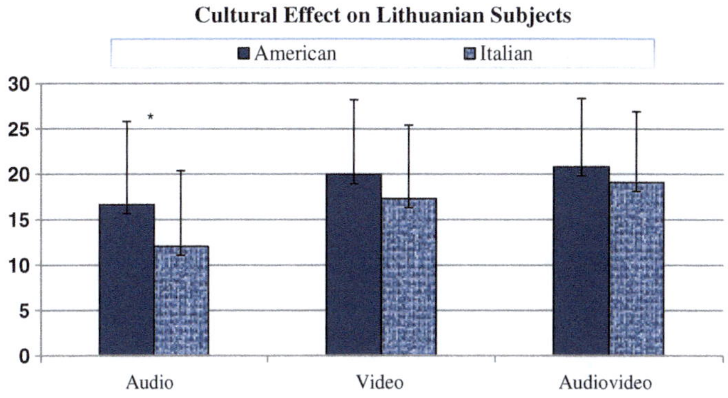

Fig. 7.3 Mean number of correct responses (and standard deviations) obtained by each 2 groups of 30 Lithuanian subjects tested on the American and Italian audio, video, and audio-video emotional stimuli, separately

The statistical analyses obtained from Lithuanian subjects revel a significant difference in the audio modality ($F(1, 8) = 17.75$, $\rho = 0.003$). The American vocal emotional expressions (m: 16.63; SD: 7) are better identified than the Italian ones (m: 12.03; SD: 6.6). No differences were found between the American and Italian mute video ($F(1, 8) = 1.95$, $\rho = 0.2$) and audio-video ($F(1, 8) = 1.504$, $\rho = 0.255$).

7.3 Comments

Results from Lithuanian subjects assessing the American and Italian data confirm that multimodality do not improve the subjects ability in inferring emotional information. Once again, for both the American and Italian emotional expressions, there are no differences in the effectiveness of the mute video and audio-video modalities in transmitting emotional information.

In perceiving emotions, in both the American and Italian expressions, Lithuanian subjects rely more on visual information (Riviello et al. 2014).

As for French and Hungarian, for Lithuanian participants, the American vocal expressions are more effective than Italian ones to infer emotional cues, whereas visual information allows recognizing emotions independently from the cultural context and the language of the stimuli.

References

Riviello, M. T., Maskeliunas, R., Kruminiene, J., & Esposito, A. (2014). Language and gender effect in decoding emotional information: A study on Lithuanian subjects. In S. Bassis et al. (Eds.) Recent *advances of neural network models and application, smart innovation, systems and technologies*, *26*, 353–61. ISSN 2190-3018. doi:10.1007/978-3-319-04129-2_35, ISBN 978-3-319-04128-5. Switzerland: Springer.

SPSS Inc. Released. (2008). SPSS statistics for windows, version 17.0. Chicago: SPSS Inc.

Chapter 8
Results for Cross-Cultural Comparison

Abstract This chapter reports the results of perceptual experiments aimed at exploring possible differences in the human ability to identify emotional information. In particular, different groups of Italian, American, French, Hungarian and Lithuanian participants were involved in the assessment of emotional expressions extracted from Italian (a country-specific language) and American English (as a global spread language) live recording movies scenes. The emotional stimuli are portrayed through the visual, auditory, and audiovisual mode. Results show that the emotional visual information share distinctive features across cultures, allowing similar emotional recognition accuracy among the different cultural groups participating to the experiments whereas the recognition accuracy of vocal emotional expressions is affected by the language specificity.

Keywords Emotional recognition · Decoding processing · Dynamic expressions · Cultural effects · Cross-Cultural experiments · Language familiarity · Cross-modal emotional database · Language specificity · Vocal emotional information · Visual emotional information

8.1 Cross-Cultural Comparison

To assess the effects of the cultural context and language, a cross-cultural comparison was made, where the performance of each participating country was separately evaluated on each American and Italian set of stimuli. Figures 8.1 and 8.2 report the comparison on the data obtained by each group of 30 American, Italian, French, Hungarian and Lithuanian subjects separately tested on the American and the Italian audio mute video and audio-video stimuli.

The ANOVA analyses (SPSS 2008) computed on the data considered the *Subjects' Nationality* as a between subjects variable and the *Emotions* and *Actor's Gender* as within subjects variables. Significance was established for $\alpha = 0.05$.

M.T. Riviello and A. Esposito, *On the Perception of Dynamic Emotional Expressions: A Cross-Cultural Comparison*, SpringerBriefs in Cognitive Computation 6, DOI 10.1007/978-94-024-0887-4_8

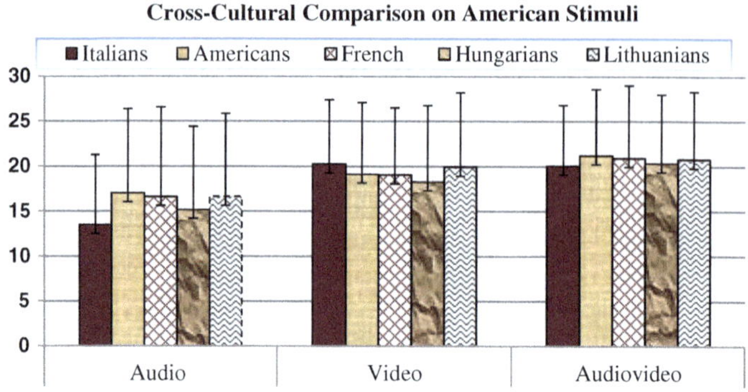

Fig. 8.1 Mean number of correct responses (and standard deviations) obtained by groups of 30 Italians, Americans, Hungarians, and Lithuanians respectively on the American audio, video, and audio-video emotional stimuli

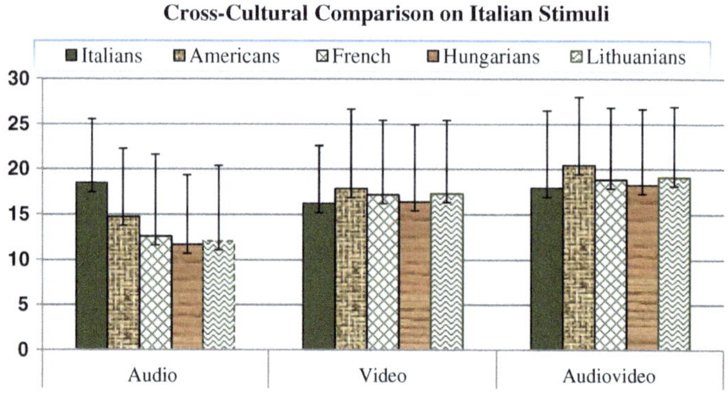

Fig. 8.2 Mean number of correct responses (and standard deviations) obtained by groups of 30 Italians, Americans, Hungarians, and Lithuanians respectively on the Italian audio, video, and audio-video emotional stimuli

The statistical results show that there were no significant differences among the data obtained by Italian, American, French, Hungarian, and Lithuanian subjects when they evaluated either the American audio ($F(4, 20) = 2.820$, $\rho = 0.53$), mute video, ($F(4, 20) = 0.548$, $\rho = 0.703$), and audio-video ($F(4, 20) = 22.79$, $\rho = 0.868$), or the Italian mute video, ($F(4, 20) = 0.771$, $\rho = 0.406$), and audio-video ($F(4, 20) = 0.971$, $\rho = 0.473$) expressions. Significant difference were found for the Italian audio stimuli ($F(4, 20) = 22.79$, $\rho = 0.000$).

Bonferroni post hoc test (Table 8.1) (Haynes 2013) revealed that American, French, Hungarian, and Lithuanian significantly differ from Italian subjects in their ability to recognize emotional information conveyed by Italian vocal expressions.

Table 8.1 Bonferroni post hoc test results

Nationality	Italians	Americans	French	Hungarians	Lithuanians	F (4,20)
M	18.40[a]	14.68[b]	12.55[b]	11.63[b]	12.03[b]	17.81***
SE	0.66	0.66	0.66	0.66	0.66	

Note Comparisons based on estimated marginal means; the mean difference is significant at the 0.05 level; adjustment for multiple comparisons: Bonferroni; a > b

***$p < 0.001$

8.2 Comments

Results from the cross-cultural comparison display how the recognition accuracy obtained from participant from all nationalities is comparable when they are tested both on American and Italian visual information, suggesting that the visual channel shares emotional features across cultures. On the other hand, the recognition of emotions from vocal information is affected by the language specificity. Indeed, no differences were found in the ability of participants (included Americans, native speakers of the language) to infer emotional information from American (a globally spread language) vocal emotional expressions, whereas American, French, Hungarian and Lithuanian participants significantly differ from Italians when tested on Italian (a country-specific language) emotional voices.

References

Haynes, W. (2013). *Bonferroni correction. Encyclopedia of systems biology* (pp. 154–154). Springer: New York. ISBN: 978-1-4419-9862-0.

SPSS Inc. Released. (2008). SPSS Statistics for windows, version 17.0. Chicago: SPSS Inc.

Chapter 9
Discussion and Conclusions

Abstract The present chapter reports the final considerations on the results of cross-cultural experiments that analyze the ability of participants from different western cultures and languages—Italian, American, French, Hungarian and Lithuanian—in inferring emotional information from audio, mute video and audio-visual emotional expressions. The emotional stimuli exploited were extracted from living recording movie scenes relating to two different cultural contexts and languages: American English (as a globally spread language) and Italian (as a country-specific language). The goal was to explore the effect of cultures and languages on the effectiveness of dynamic visual and vocal cues in conveying emotional expressions. The main findings suggest that the processing of dynamic visual information shares more commonalities trough different close cultures, whereas the identification of vocal emotional information seems to be more affected by the language familiarity. In addition, a language cultural specificity as well as a nonlinear processing of the emotional information is hypothesized. A discussion on the recognition accuracy of some of basic emotion is also reported.

Keywords Cross-Cultural experiments · Language familiarity · Emotional database · Cultural specificity · Emotional recognition · Dynamic expressions · Cultural effects · Emotion categories · Auditory emotional information · Visual emotional information

The present work focused on perceptual experiments, involving separate groups of Italian, American, French, Hungarian, and Lithuanian participants, with the aim to explore the ability to recognize emotional expressions dynamically portrayed through the visual and auditory channel across close cultures.

The main goals were to investigate if one channel is more effective than the other to infer emotional information and if this effectiveness is affected by the familiarity with the cultural context and in particular with the language.

To this aim, the perceptual experiments rested on the perception of dynamic visual and vocal cues exploiting two databases of audio, mute video and audio-video emotional expressions based on video-clips extracted from American

M.T. Riviello and A. Esposito, *On the Perception of Dynamic Emotional Expressions: A Cross-Cultural Comparison*, SpringerBriefs in Cognitive Computation 6, DOI 10.1007/978-94-024-0887-4_9

and Italian movies, respectively. The rationale was to provide emotional stimuli arisen from a very spread and known culture and language, as American is, and a country specific cultural and linguistic background as the Italian one.

In that way, the databases allow the analysis of audio and video emotional expressions with the aims to define and identify distinctive emotional features from multi-modal and cross-cultural signals as well as for developing new methodologies and mathematical models for the automatic implementation of naturally human-like communication interfaces.

The results first show that in perceiving emotions, expressive information is not added over the amount of emotional cues provided. For participants of all nationalities, the bimodal presentation of the emotional information (audio-video combined modality) did not significantly improve subjects' recognition accuracy for both American and Italian stimuli, suggesting a non-linear processing of the emotional features. In particular, for American, French, Hungarian, and Lithuanian participants the audio-video conveyed the same amount of emotional information conveyed by the mute video mode, suggesting that dynamic facial expressions perform a robust encoding of the emotional features, and seem to be more effective than the vocal ones in communicating emotions.

This result is challenged by a language cultural specificity: the Italian subjects when tested on emotional stimuli extracted from Italian movies have similar recognition performance in the audio alone and the combined audio-video modes, showing that they rely more on the vocal information as native speakers of the Italian language. Since this is not true for American participants, who favored visual information in perceiving emotions through American stimuli, it seems not to be the language expertise to allow a better identification of emotional cues through vocal expressions. This result may be due to an effect of the video-clips used as stimuli, but may also bring to the hypothesis that speakers of different languages exhibit a different sensitivity to vocal emotional information that could be attributed to the language structure: i.e. different languages weight differently the amount of emotional information encoded in their supra-segmental structure.

The effectiveness of the visual channel in communicating emotions is further supported by the results of the cross-cultural comparison showing no significant differences in the recognition performance of the five country's participants on both American and Italian mute video and combined audio and video stimuli, suggesting that the processing of visual information shares more commonalities trough different close cultures, independently from the cultural context and language through which the emotional expression is conveyed. The identification of vocal emotional information, seems to be more affected by the language familiarity since, there were no significant differences when participants assessed American emotional vocal expressions (globally spread and second language used by the involved subjects) whereas, this was not true for the Italian emotional vocal expressions: non-native speakers' participants significantly differ from Italians.

In addition, French, Hungarian and Lithuanian participants perform better on the American than on the Italian emotional voices, supporting the language specific effect hypothesis suggested above.

Regarding the emotion categories under examination, it has been shown that among the primary emotions, anger is perceptually privileged since, regardless of the cultural context and communication mode, it obtained, by all country participants, the higher decoding accuracy.

Assuming that the experimental set up proposed is testing the capability of the subjects to infer emotional cues from the interlocutor, identifying anger may activate cognitive self-defense mechanisms that are crucial for the perceiver's survival, hence humans may have a high sensitivity to recognize it independently from the cultural environment. Among the remaining emotional categories under consideration, happiness is more easily identified through the visual information both for American and Italian stimuli.

As expected, non-native speakers poorly recognize both American and Italian ironic expressions, probably because the perceptual cues of this emotion are strictly linked to the language and cultural context. Surprise is definitely the emotion less well recognized among the others. It could be hypothesized that this emotion represent only the first, brief stage of an emotional state which often evolve in another one (fear, happiness, etc.), challenging the inclusion of surprise in the list of the "basic" emotions.

The reported research work is an attempt to go into behavioral mechanisms characterizing the human decoding process of emotional features exploiting dynamic, realistic and mutually related vocal and visual information.

More data are needed to support the hypotheses developed and described above, for example, by extending the proposed perceptual experiments to the members of other Western and non-Western countries, as well as by defining and exploiting cross-modal databases extracted from different cultural and linguistic context.

Some considerations for improving the present work's outcome can also be made. Taking advantage of the huge number of participants testing the proposed databases, a re-definition of the collected stimuli can be planned accounting for the subjects' agreement on each stimulus labeling. This could lead to a more effective collection of emotional data to be exploited for future perceptual and automatic recognition task.

As the above results disclose it would be worth performing a more accurate analysis of the contribution of each specific emotion to the subjects' perceptual performance, i.e. the inter-relationships among the emotion categories, the effectiveness of communicative channels (visual and auditory) and the cultural and linguistic background of the stimuli and the participants.

Finally, among the aspects related to human ability in perceiving emotions, it would be interesting investigate the role of the subjects' and the actors' gender in identifying and portraying emotional cues and the possible interrelations between the two factors.

Index

M.T. Riviello and A. Esposito, *On the Perception of Dynamic Emotional Expressions: A Cross-Cultural Comparison*, SpringerBriefs in Cognitive Computation 6, DOI 10.1007/978-94-024-0887-4

GPSR Compliance

The European Union's (EU) General Product Safety Regulation (GPSR) is a set of rules that requires consumer products to be safe and our obligations to ensure this.

If you have any concerns about our products, you can contact us on ProductSafety@springernature.com

In case Publisher is established outside the EU, the EU authorized representative is:

Springer Nature Customer Service Center GmbH
Europaplatz 3
69115 Heidelberg, Germany

Batch number: 10844675

Printed by Printforce, the Netherlands